YOUR KNOWLEDGE HAS VALUE

- We will publish your bachelor's and
  master's thesis, essays and papers

- Your own eBook and book -
  sold worldwide in all relevant shops

- Earn money with each sale

Upload your text at www.GRIN.com
and publish for free

Bibliographic information published by the German National Library:

The German National Library lists this publication in the National Bibliography; detailed bibliographic data are available on the Internet at http://dnb.dnb.de .

This book is copyright material and must not be copied, reproduced, transferred, distributed, leased, licensed or publicly performed or used in any way except as specifically permitted in writing by the publishers, as allowed under the terms and conditions under which it was purchased or as strictly permitted by applicable copyright law. Any unauthorized distribution or use of this text may be a direct infringement of the author s and publisher s rights and those responsible may be liable in law accordingly.

Imprint:

Copyright © 2015 GRIN Verlag
Print and binding: Books on Demand GmbH, Norderstedt Germany
ISBN: 9783668729117

This book at GRIN:

https://www.grin.com/document/429304

Amos Wesonga

Prevention throughout the lifespan. Mental health in adolescence

GRIN Verlag

GRIN - Your knowledge has value

Since its foundation in 1998, GRIN has specialized in publishing academic texts by students, college teachers and other academics as e-book and printed book. The website www.grin.com is an ideal platform for presenting term papers, final papers, scientific essays, dissertations and specialist books.

Visit us on the internet:

http://www.grin.com/

http://www.facebook.com/grincom

http://www.twitter.com/grin_com

Table of Contents

- Part 1 2
  - Introduction 2
  - Importance of prevention 2
  - Recommendations 3
  - References 4
- Part 2 5
  - Local community prevention program for the Young Adult life stage 5
- Part 3 6
  - Thesis statement 6
  - Alcoholism has the following effect on workplace 6
  - Prevention of alcohol in workplace 7
  - Alcohol's Effect on Family 7
  - Culture and subculture 7
  - Impact on the Community 7
  - Conclusion 8
  - References 9

Part 1

Introduction

Adolescence is the transition from childhood to adulthood. Every Child's social and emotional development is different. Kid's development is a product of his or her unique combination of development of the brain genes, environment, involvement with family and friends, influence from the community and culture. Mental health is vital during adolescent since it includes emotional, psychological, and social well-being. Also, it affects how a person thinks, feel, act, handle stress, relate to others, and make choices. Adolescence involves many changes, both positive and negative. There is need to prevent the negative ones from occurring since they may cause adverse effects on one's life.

Importance of prevention

During the adolescent stage, children are in developmental transition. They are sensitive to environmental or surrounding influences. Some environmental factors, such as family, peer group, school, neighborhood, policies, and societal cues can either support or challenge young people's health or well-being. Ensuring that they have good mental health helps them to take important decisions in life and lead a better adult life (Saxena, 2012). Prevention at this stage is critical, given that public health and social problems such as suicide, smoking, substance use and abuse, teen and unplanned pregnancies among others begin at this stage.

Self-management behaviors initiated in adolescence are long-lasting. Family bonding plays an imperative role in prevention since parents can train their children. Programs designed by schools and community on prevention can intervene in the life of a child and transform it for the better (Cavaleri, Olin, Kim, Hoagwood, & Burns, 2014). Additionally, promotion of students' health by schools should be an essential part of its core businesses since it increases academic attainment and development in life.

Gladstone, Beardslee, & O'Connor (2011) established depression to be a common psychiatric problem among the adolescents. Major depressive episode (MDD) is approximately 2% for children and increases to 4% to 7% during adolescence. Data provided by the National Comorbidity Survey (NCS) shows that the prevalence of MDD approximate 14% for adolescents aged 18 years and 20% of the youths tends to have a depressive disorder when they reach the age of eighteen years (Eija & Kaija, 2010). Therefore, a preventive measure is the most effective approach that can help to control the rising rate of depression as children approach adolescent age.

Recommendations
Primary level
Primary level children spend more time in schools even than they do at home. Fazel, Hoagwood, Stephen, & Tamsin (2014) suggest that schools play an essential role in the development of children in peer relationships, social interactions, academic achievement, cognitive development, emotional control, behavioral potentials and moral growth. Therefore, learning institutions should create a range of integrative care that can progress mental health, as well as educational attainment for kids. However, this process requires a reconfiguration of education and mental health systems to implement evidence-based practices required.

Secondary level
During the secondary level, some of the mental disorders already manifest (Sareceno, 2014). Therefore, the state should aim at lowering the rate at which disorders show among the youths by detecting and treating any diagnosable conditions.

Tertiary level
Any individual who approaches the tertiary level has developed disability conditions (Sareceno, 2014). Therefore, the primary aim should be an enhancement of rehabilitation of the affected persons. At this stage, it is also worthy to prevent relapses and recurrence of illness.

References

Cavaleri, A. M., Olin, S. S., Kim, A., Hoagwood, E. K., & Burns, J. B. (2014). Family Support in Prevention Programs for Children at Risk for Emotional/Behavioral Problems. *HHS Public Access*, 399-412.

Eija, S., & Kaija, A.-S. (2010). *Mental Health Promotion in Young People - an Investment for the Future.* Copenhagen: World Health Organization.

Fazel, M., Hoagwood, K., Stephen, S., & Tamsin, F. (2014). Mental Health Interventions in School 1: Mental Intervention in Schools in High-Income Countries. *HHS Public Access*, 377-387.

Gladstone, R. G., Beardslee, R. W., & O'Connor, E. E. (2011). The Prevention of Adolescent Depression. *HHS Public Access*, 35-52.

Sareceno, B. (2014). *Prevention of Mental Disorders: Effective Interventions and Policy Options.* Geneva: World Health Organization.

Saxena, S. (2012). *Prevention and Promotion in Mental Health.* Geneva: WHO.

Part 2

Local community prevention program for the Young Adult life stage

The program explored is the "Prevention and Early Intervention in Mental Health-Puberty to Early Adulthood" used in the United States. The aim is to reduce the prevalence of mental disorders among young adults in the US through enhancing a healthy living. The foundation of this approach lies on the fact that brain disorders are a product of brain changes combined with friendship, social roles, self-confidence, hormones and the challenges faced in life. These factors work together to increase the transition and new stressors that deteriorate the mental health of teens.

The program targets the health, safety, resources, relationships, and interventions. For the health sphere of a youth's life, the program uses ways that can enhance sleep and physical activities while discouraging substance use. For safety purposes, the program aims at giving the youths the ability to evade intimate partner violence. Romantic relationships are common and essential part of the teenage life but lead to violence such as physical, emotional, psychological, and sexual abuse. The ability to deal or avoid such effects can promote the mental health of the youths. For the case of the resources, this program aims at supporting children with fears of talking about their psychological health or seeking help. Another sphere touched by this program is the relationship, and it deals with prevention of bullying and isolation. The final phase is to manage unique challenges that youths face as they transit to adulthood. The program is holistic since it touches all spheres of life of a teenager. Therefore, it can promote healthy living among the youths.

Part 3

Thesis statement

Alcoholism among young adults is a significant problem in the modern society as it causes social, health and financial challenges. Drug abuse can cause considerable fiscal issues within a short time, and health problems later in life. Excessive alcohol consumption affects all organs of the body- the liver, the immune system, and heart. Driving when drunk is a primary source of road accidents. Therefore, alcohol consumption does not cause any benefit to the life of a young parent.

Alcoholism has the following effect on workplace

Consumption of alcohol leads to misuse of the earned money and later may lead to loss of valuable skills. Also, it increases the risk of diabetes, stroke, heart failure, lung disease, cirrhosis of the liver, HIV/AIDS, hepatitis, and tuberculosis (Bender, Leone, Szumski, & Huagen, 2014). It also causes cancer and affects the digestive system. Estimated billion is spent each year on catering for treatment of alcohol-related injuries. As such, it channels hard-earned money to unnecessary use. In the long run, the patient may succumb to death, leading to considerable loss of valuable expertise.

Consumption of alcohol causes poor concentration, lack of focus, carelessness, and errors in judgment. A combination of these effects reduces productivity by even one-third since it increases the rate of absenteeism (Klingemann, 2015). Employees who are alcohol addicts are more likely to have injury-related issues, sleep on the job, and struggle to concentrate. They are more likely to spend more working hours on consuming alcohol or other drugs.

Alcohol can cause adverse changes in a person's behavior and may lower the level of self-esteem. For instance, the addict may become weary, self-justifying, or quick-tempered. Alcohol addicts may also suffer from stress due to financial problems, complain about issues at home, or blame their coworkers for their mistakes and shortcomings. They also tend to neglect their hygiene and appearance, and they may portray symptoms of a hangover or withdrawal. The motivation of the entire workplace is likely to be affected by the behavior of alcohol addicts. For instance, other employees may be forced to chip in the duties left by the addict when absent or late, and this may lower the productivity at work.

Prevention of alcohol in workplace

The best way to handle the problem of alcoholism in the workplace is to promote ideas that employers can monitor drug and alcohol abuse. These may include rehabilitation services for employees, seminars on alcohol consumption, and implementing strict rules that discourage the use of drugs. Employers are also recommended to look into Employee Assistance Program (EAP) vendors, which offer a variety of services to workers such as financial advice, personal counseling, and referrals to addiction services (Azzine, 2009). EAPs are intended to help employees make wise choices if a coworker or a family member has a problem with alcohol, prescription or illicit drugs. Their services help employees to remain productive.

Alcohol's Effect on Family

Children growing up with parents who abuse alcohol are likely to develop alcohol use disorder later in their lives. Addict parents are likely to neglect their kids and when they are high on drugs, may injure their offspring, especially when they get angry. In extreme cases, the state may be forced to separate such a parent from his children. The drug affects an individual's cognitive functions and physical capabilities. As a result, addicts may neglect their responsibilities in the workplace and home, leading to low-income generation, which may eventually challenge one's capacity to provide basic needs to the family. In extreme cases, marriage is at stake. Useful prevention measure can help to ensure that every family member is in a position to lead a healthy life under all circumstances.

Culture and subculture

In some cultures, responsible drinking is allowed. For instance, some cultures only permit the elderly to consume alcohol while the young are not allowed. Drinking habits may also be combined with accepted norms and regulations concerning who may drink, how much, when, how, the contexts, and the effects of other limits (Moore & Cissar, 2009). Some of these cultures hinder the practical process of controlling alcohol consumption in some communities. As a result, alcohol-related mental disorders are prevalent in such communities.

Impact on the Community

Mental disorders related to alcohol are prevalent in the modern communities. Moss (2013) notes that heavy drinking has a close relationship with the increased level of crime, violence, and unemployment. As a result, a community whose population is affected by alcoholism does not advance economically. A report by WHO (2008) indicates that work performance had a negative correlation to the volume and drinking pattern. The main reason for the existence of such a relation is because drug abuse leads to lower performance, lack of self-direction and social problems.

In most cases, communities have tried to fight the problem through some measures such as rehabilitation services. Governments and non-governmental organizations have also played a significant role in providing social support services to communities with the aim of reducing alcohol consumption. However, all these initiatives have not succeeded in reducing the level of alcohol intake. Other approaches used is the advertisement of negative impacts of alcohol through the media. The major problem with the initiatives used is that they try to deal with the effects of alcohol, instead of the cause of the problem. As a result, mental disorders related to alcoholism continue to affect the community, an impact that may continue if sustainable measures are not taken. The following recommendations can be useful in controlling the level of alcohol consumption.

i. Discourage the youths from engaging in alcohol consumption behaviors
ii. Promote healthy and responsible living among the children
iii. Implementation of strict rule and regulations to control the production and sale of alcohol.

Conclusion

From the above analysis, it is clear that alcoholism has promoted mental disorders among the young generation. As a result, their productive capacity has gone down significantly. Their economic conditions have also deteriorated. As a result, the ability to support their families has declined substantially. The problem has continued to affect the society due to inadequate control measures used. As a result, the study recommends the use of active measures such as dealing with the cause of the dilemma. For instance, preventing children and adolescents from engaging in alcohol drinking may reduce the related mental disorders effectively in the future.

References

Azzine, V. (2009). Workplace Stress, Organizational Factors, and EAP Utilization. *Journal of Workplace Behavioral Health*, 344-356.

Bender, D., Leone, B., Szumski, B., & Huagen, M. D. (2014). *Alcoholism* . New York: Current Controversies.

Klingemann, H. (2015). *Alcohol and its Social Consequences - The Forgotten Dimension.* Geneva: World Health Organization.

Moore, W., & Cissar, D. (2009). *A Healthy Drinking Culture: A Search and Review of International and New Zealand Literature.* Alcohol Advisory Council of New Zealand.

Moss, H. (2013). The Impact of Alcohol on Society: A Brief Overview. *Journal of Public Health*, 175-177.

WHO. (2008). *Global Status Report on Alcohol 2004*. Geneva: World Health Organization.

YOUR KNOWLEDGE HAS VALUE

- We will publish your bachelor's and master's thesis, essays and papers

- Your own eBook and book - sold worldwide in all relevant shops

- Earn money with each sale

Upload your text at www.GRIN.com
and publish for free